YOU WILL GO

 Published in New York by Beginner Books, Inc., and simultaneously in Toronto, Canada, by Random House of Canada, Limited. Library of Congress Catalog Card Number: 59-6469. Manufactured in the United States of America.

BOOK CLUB EDITION

TO THE MOON

MAE and IRA FREEMAN

Illustrated by ROBERT PATTERSON

Beginner Books
A DIVISION OF RANDOM HOUSE

The moon is up there,

far away.

No one has been there yet.

But some one will go there soon.

Some day you may go there, too.

This is how you will go
to the moon.
Here is the rocket
that will take you up into space.

It is a tall, tall rocket.

It is as tall as ten houses.

The rocket has 3 parts.

You will go way, way up to Part 1.

The rocket men will take you up.

They will take you up

in a little car.

Come on in.

Come into this little room.

This is where you will sit.
You will sit here
with the rocket men.

The men will show you
what to do.
The men will show you
where to sit.
Hook on that belt.
Hook it on tight!
Get set to go!

TV
TV

BOOM!

At last!

The rocket is off!

And you are in it!

3
2
1

You feel the rocket shake!

You feel the push. Push. PUSH!

But you do not fall

from your seat.

The seat belt holds you tight.

Now you would like to see

where you are.

You would like to look out.

How can you do that?

Is there some way to look out?

Yes, there is a way.

TV

Turn on the TV.

Now you can see where you are.

Your rocket is high in the sky.

How black the sky looks out here!

What is that funny little thing

out there so far away?

It is the space station.

Your rocket will go there.

BUT . . .

Not ALL of your rocket

will go there.

The fuel in Part 3 is now all gone.

Your rocket dumps Part 3.

Part 3 falls away.

Now Part 2 fires.

On and on into the sky

you will go.

Soon Part 2 has no more fuel.

Now your rocket must dump Part 2.

It falls away, too.

Now there is just Part 1,

AND YOU ARE IN IT!

Soon you get near
the space station.
Now you can see
that it looks like a big,
big ring.

And you can see the space station turn.
It keeps turning all the time.
You will stop at this space station
on the way to the moon.

There are men here
at the space station.
They will fit your rocket
into the wall.
You will have to
get out of the rocket here.

Open your belt.

OH! Up you go.

You can not stay down!

You do not fall!

YOU ARE A SPACE MAN NOW.

At home all things stay down.
The pull of the earth
holds them down.
But not one thing will stay down
out in space.

Come into the space station now.

Come to the ring part of the space station.

Things DO stay down here.

You will stay down too!

Why?

The space station turns and turns.

The turning is what keeps things down.

So it is more like home here.

There are things to do here.

You can eat.

You can sleep.

You can see a show.

How long will you be here?

When will you go on

to the moon?

Soon.

You will go as soon as the moon ship

comes for you.

Look! What is that funny thing out there?

It is the moon ship.
It does not look like a ship at all.
It looks like a big bug
with bent legs.
The moon ship will fit into the wall
of the space station.
Then you can go into the moon ship.

AIR

Now you are in the moon ship.
The moon ship is not
like the space station.
You can not keep down.

Hold on to the rings.
That is how you will
get to your seat.

Now the rockets fire.

You feel the big push.

Your moon ship gets away fast.

THEN . . . .

The rockets stop!

They gave

your moon ship a big push.

That one big push will make it

go on and on.

Now there is not one thing to hold your moon ship back.

There is not even any air to hold it back.

There is no air at all out in space.

There will be no more push.

So you can open your belt now.

You will be in this moon ship
for three days.
That is how long it will take
to go to the moon.
Here are things to eat.
Here are books and games.

Here is TV, too.
You can get stations
from earth on this TV!
There are many things
for you to do.
How fast the three days go!

At last! This is the big day.

Look! There is the moon.

How big it is!

It is like a big ball.

It has holes and bumps all over it.

How will the moon ship land?

A little rocket fires.

It turns the moon ship.

Now the moon ship can land

on its legs.

Get set to land!

Now the BIG rockets fire.
They hold the moon ship back.
So there will be no bump
when you land on the moon.

Down you go.

The legs of your moon ship fold out.

The moon ship lands on its legs

like a big bug.

YOU ARE ON THE MOON!

But you can not go out yet.
There is no air for you out there.
You could not live
in the hot, hot days on the moon.
You could not live
in the cold, cold nights on the moon.
But put on this space suit,
and then you can live on the moon.

See the little radio station
in your space suit!
It will let you hear
what the others say.
And you can say things to them, too.

The moon does not look like the earth at all.
There are no trees!
No lakes! No water at all!
There is just deep, gray dust.
Dust, dust, dust.
There is a big hole over there
that looks like a dish.

There are many dish holes
all over the moon.
Look over there! What do you see?
Those funny things are moon cars.
One of them will take you for a ride.

Away you go in the moon car.
It will take you to those high hills.

A big dish hole is in the way.
But a moon car does not turn away
from a dish hole.
It can go right down into the hole.
Down, down, down.
Then up, up, up. And out.
And away you go to the hills.
You will get out of the moon car there.

You will get out of the car
and have some fun.

Say, look what you can do!
You can get to the hill
in five big steps.

You can climb to the top
of this high hill on the moon.
You can climb to the top fast!
You could not climb
this kind of hill at home.
But you can here!

The moon does not pull you down

as much as the earth does.

See! Up you go!

You can see more from the top of this hill.

Look! Do you see that house?

That is the moon house.

That is where you will live on the moon.

Now look out at the sky.
Do you see that big dot up there?
That is Mars.
Mars is a long, long way
from the moon.
What would you find on Mars?

No one knows yet.

But some day you may go there, too!

Then YOU will see.

The Science of Your MOON TRIP

Now you are back from your trip to the moon. You can tell your friends about it. When you tell them, you can use some of the words that scientists use.

You went to the moon in a rocket. Rockets and airplanes are not the same at all. An airplane must push on the air around it. It can not fly where there is no air. But a rocket can go even in space, where there is no air. When rocket fuel burns, gas shoots out in back. This kicks the rocket ahead. Scientists say the rocket works by **REACTION** (*See pages 12, 36, 44, 46*)

When you were out in space, the sky looked black. The sun did not light up the sky. You could see the stars in the daytime! In outer space there is no air to spread the rays of the sun. The sky looks black all the time. Near the earth there is air to catch the sunlight and throw it all around. Then the sky lights up and looks blue. This is called **SCATTERING OF LIGHT** (*See page 16*)

The earth pulls everything toward itself. This makes things stay down. The moon pulls things toward itself, too. But the moon is smaller than the earth, so it does not pull as much. That is why it was so easy to climb a high hill on the moon. That is why you could take such giant steps on the moon. The pull of the earth or moon is called **GRAVITY** (*See pages 27, 55, 56*)

Things have no weight out in space because there is so little gravity. There is nothing to hold them in place. That is why you floated around in the rocket and in the moon ship. That is why everything was clamped down. Scientists say that things out in space are **WEIGHTLESS** (*See pages 26, 35*)

When you went into the space station, it seemed like home again. There was no gravity but you could stay down! Why? The turning of the space station made things pull out toward the edge of the ring. This pull worked like gravity and kept things down. An outward pull like this is called **CENTRIFUGAL FORCE** (*See page 29*)

Nothing starts to go all by itself. Everything stays where it is put unless something else MAKES it move. The rockets of the moon ship had to push to start it moving out into space. After anything is started, it keeps moving if there is nothing to stop it. Your moon ship kept going even after the rockets were shut off. That is the way it is with ALL things. If they are still, they try to stay still. If they are moving, they try to keep moving. Scientists call this **INERTIA** (*See pages 14, 37, 38*)

Sounds can go only where there is air to carry them. But radio can go even where there is no air. That is why there was a little radio station in your space suit. That is how you talked to the other people when you were on the moon. Engineers say that you used **RADIO COMMUNICATION** (*See page 48*)

A dish hole on the moon is called a **CRATER** (*See pages 51, 53*)